S.S. 163.]   O.B. 2025.

HINTS ON RECONNAISSANCE FOR MINES AND LAND MINES IN THE AREA EVACUATED BY THE GERMANS.

(Compiled chiefly from Notes forwarded by Inspector of Mines.)

The greatest care must be taken when reconnoitring ground, evacuated by the Germans, for mines and land mines.

A cursory examination is not sufficient, as the mines are often extremely well concealed.

The following suggestions, compiled from various sources, may be of use to Officers making these reconnaissances:—

Cross-roads.

Cross-roads are often mined, and the charges have been known to explode as long as 48 days after our occupation. The surface of the road should be carefully examined for signs of fresh work. Any found should be at once opened out to ensure that there is no contact mine there. (Sketches of typical examples—Nos. 1, 2, 3.)

The ground should be cleared for 25 yards in every direction from the cross-roads, and carefully inspected for signs of the entrance to any gallery which may pass under the road.

The entrance may be blown in, or covered in by the *débris* from another explosion, after the charge has been laid.

Any small craters should, therefore, be carefully investigated.

The following examples have been found:—

1. In Noyon, galleries had been driven from under the paving stones.

2. Gallery driven under road, and excavated till only a thin crust remained. 8-in. shell in position with the fuze portion on, but loose and in contact with the roof.

3. Shell-holes on a road, apparently filled up with bricks, etc., have been found to contain land mines fitted with instantaneous fuze and friction fuze lighter, with safety pin attached to trip wire (which had broken off).

Dug-outs.

Any dug-outs left undestroyed should be carefully examined.

Look for any setts from which the tenons have been cut and replaced by wedges.

Loose boards in floor, sides, or roof often locate the smaller variety of charges.

Apparent dead-ends should always be inspected carefully; the two biggest charges were found behind such places, complete even to the pick marks one would naturally expect.

Every alternate sett should be removed and replaced.

The following examples of what have been found may serve as a guide to what may be expected:—

1. Grenades liable to explode when trodden on.

2. Dozen stick-bombs, arranged to be fired by wire attached to sandbag which had to be removed to open a dug-out door.

3. Wires to fire charge attached to handrail in entrance of dug-out. (In dug-outs constructed with casing, mortice and tenon joints, the position of a charge is sometimes indicated by the wedging of the timber, where sides have been cut and removed.)

4. Charges of 2,000 lbs. with 20 ft. of tamping in wall of dug-out connected with a pair of firing leads amongst a number of telephone wires. (Intended method of firing not discovered.)

5. In two cases, charges of a few hundred pounds left in undamaged dug-outs, which were attractively equipped so as to induce early occupation, exploded about 8 days after enemy evacuation, presumably by clockwork or other delay-action device.

6. A shovel stuck in the side of a dug-out between timbers. The wires from battery of three

dry cells were, one attached to metal and the other to contact plate. The shovel stuck out as an obstruction, and would in the ordinary way have been removed.

7. A false step in the stairway of dug-out of thin planking making contact when trodden on.

8. A blown-in entrance to a dug-out is not always a safety sign. Charges may be concealed in the unblown portion. They are generally crudely arranged contact charges.

9. An elaborate and neat trap has been discovered under the Achiet-le-Grand—Bapaume railway embankment, S.E. of Bihucourt. Above the timber ceiling of a tunnelled stairway leading to dug-outs under the embankment was a mine to which access was obtainable only by removing three of the setts (frames) of the stairway. Every sett was intact and carefully wedged, there were no gaps, and the hand rails were continuous. (Sketch 4.)

10. A window weight, suspended by a fine cord crossing entrance, arranged to drop into a box of detonators in connection with charge.

11. One of the pieces of timber on the side of the stairs leading down into a dug-out projected slightly inwards at the top, though it was in place at the bottom. It was pulled out for investigation, and a nail was found driven through its lower end, the point of which was placed against the cap of a cartridge which had a charge of explosive behind it. Had the plank been forced into its correct position, the nail would have struck the cap and exploded the charge.

12. Branch placed over entrance to dug-out as if to conceal it, when removed caused an explosion *2 minutes later*, completely destroying dug-out.

13. Trip wires in entrances of dug-outs, etc., arranged to explode charges or grenades.

14. Charges have been found:—In chambers on each side of the entrance; in chambers off the dug-out itself; in the ventilating shafts.

15. Charges are usually found to be 80–150 lbs. of Perdit placed in small chambers at a height of 5 ft. from the floor, and in the ventilating shafts 10 ft. or 11 ft. below ground level, and in every case at the end of a little gallery 4 ft. long by 18 in. square. The charge is tamped with a wooden panel backed by loose stones cemented over at the end. Firing is electric by armoured cable.

Houses.

A house of any size left standing should always be looked upon with suspicion. The cellar especially should be carefully examined and the surface inspected and ground around the house cleared of debris, as mines are sometimes sunk against the wall of cellar.

The following examples have been found:—

1. A box of explosives buried in a cellar, timed to go off by the corrosive action of acid on a steel wire. (Sketch 7.)

2. Charges, with fuze and detonator, in chimney.

3. Detonators in lumps of coal.

4. Book on table, with wire down leg of table. Charge would fire if book were lifted.

5. A mechanical fuze igniter attached by wires to an explosive charge fixed in the walls of a house has been found in Neuville-Bourjonval.

6. In the paving of a house in Roye, the Germans had sunk a hole from the ground floor to a stone drain five metres below the ground level; under this drain there was a concealed shallow well; from this they had driven out two small galleries and charged them each with 150 lbs. Perdit. The hole to the drain had been filled in again and repaved.

7. The French experimented successfully for clockwork devices by means of the Geophone.

8. Grenade under loose brick in floor of stable covered with straw. Pressure on brick would explode grenade.

Railways.

Especial care should be taken with the investigation of the following places for signs of enemy work :—

Bridges.—Charges are often placed on the girders, or holes are sunk in the abutments behind the girders.

The approaches to a bridge which has been destroyed should be examined. Trap charges have been discovered which were laid with a view of destroying the temporary structure over the gap.

Level Crossings.—In some cases mines have been driven under the crossing by means of an inclined gallery from the flank.

Embankments.—Charges have been discovered at the ends of galleries, driven into the embankment. A land mine was discovered fixed inside a rectangular box 8 in. × 8 in. section, 10 ft. long. This box was sunk vertically in the embankment between the rails. One foot of earth was rammed in on top of the explosive, which was to be fired by electrical means.

Wells.

Wells are often destroyed by boring a 6 in. hole, 10 ft.-20 ft. deep, a few feet to one side of the well, filling this with explosive and blowing it. (Sketches 5 and 6.) The following tips for locating wells blown in this manner may be of use.

1. The German sign for a well is a white board bearing a red ring with a red disc in centre, or the word "BRÜNNEN."

2. The locality of a well is usually a crater in the yard of a building. The charges are usually placed about 12 ft. from side of well, 10 ft.-20 ft. deep, so that the well is never in the centre of crater.

3. Men employed reclaiming a well should work with a life line on. Novita Sets should be handy in case of any men becoming gassed from fumes of explosive which was burnt.

4. Wells and ponds have been rendered unfit for drinking by means of Creosol, dung and all sorts of filth. Wells should be labelled "NOT TO BE USED," until the water has been tested by the local expert.

Bathing Places.

Pointed stakes have been found driven in with their points below water level, and interlaced with barbed wire.

Bathing places should be examined *before being taken into use by the troops.*

General.

The following remarks may be useful as regards the search for and destruction of an enemy's land mines. Great skill and care are required. Suspected localities should first of all be studied with good field glasses. The following signs should be looked for,—freshly turned up earth, settlement of the ground, oval marks on the ground after rain, patches of grass that stand out conspicuously, narrow strips where the earth has been disturbed which may mark where leads have been laid, ends of wire, cord and canvas sticking up, numerous foot tracks on a confined space, litter of materials, such as powder, guncotton, shavings, paper. Suspicious places in soft ground can be investigated with a probe. If a contact mine is discovered, it should be marked and destroyed later by firing a slab of guncotton on top of it. Trip wire mines can be destroyed by attaching a guncotton primer to the wire and detonating it, or by firing the mine by means of a long cord made fast to the trip wire. When the leads of observation mines are discovered they should be cut singly and the ends turned up. Contact mines have sometimes been exploded by driving cattle over them. The following devices have also been discovered:—

1. Barricades interlaced with wires attached to stick-bombs.

2. Hand-bombs buried in trench with telephone wires attached.

3. Trench boards, new in every case, on several fire steps which detonated bombs when trodden on.

4. 7 in. shells with fuze removed and replaced by detonator.

5. Cap badges, artificial flowers, bits of evergreen, pieces of shell and other articles likely to be picked up as "souvenirs," attached to charges.

6. The preparations for blowing up Fort de Conde appear to have involved charges of 2,600–3,000 lbs. to be fired independently after the fort was captured. The electric leads were duplicated—one being apparent and the other buried $1\frac{1}{2}$ ft. to 2 ft. below it.

Explosives.

The explosives used by the Germans are Westphalite, Perdit and Donarit. They are all hygroscopic. Charges found by the troops may therefore be rendered reasonably safe in the first instance, by being saturated with water. They should be left *in situ* to be removed by men accustomed to handle explosives.

The withdrawal of charges must be done with care, as detonators are frequently found distributed throughout them. This is specially the case with portable charges made up in tins. Detonators have been found in the middle packets of one of these. Each charge should be opened for examination.

Exploration for charges leads to the accumulation of a large amount of loose explosive, which is unfit to return to store. This should be destroyed as soon as possible in one of the following ways:—

    1. By detonating it in small quantities, in consultation with troops in the vicinity.

    2. By scattering it broadcast over waste ground. It should be remembered that if the ground is subsequently occupied by animals as picket lines, they run some risk of being poisoned.

    3. By burning, the explosive being laid out in long parallel lines about 6 in. high. This is the best and safest way, provided all detonators have been removed. The fumes are unpleasant, and the explosive should be burnt after consultation with the troops in the neighbourhood, and carried out when the wind is in the direction which will cause least inconvenience.

    4. Ignition of the explosive is facilitated by mixing with it a little cordite or the charge of German howitzers.

Charges must not be destroyed by throwing them into ponds or down wells.

SKETCH No 7.
GERMAN AUTOMATIC DETONATING DEVICE
USED IN CONNECTION WITH EXPLOSIVE CHARGES LEFT IN DUG-OUTS, BILLETS & ELSEWHERE.

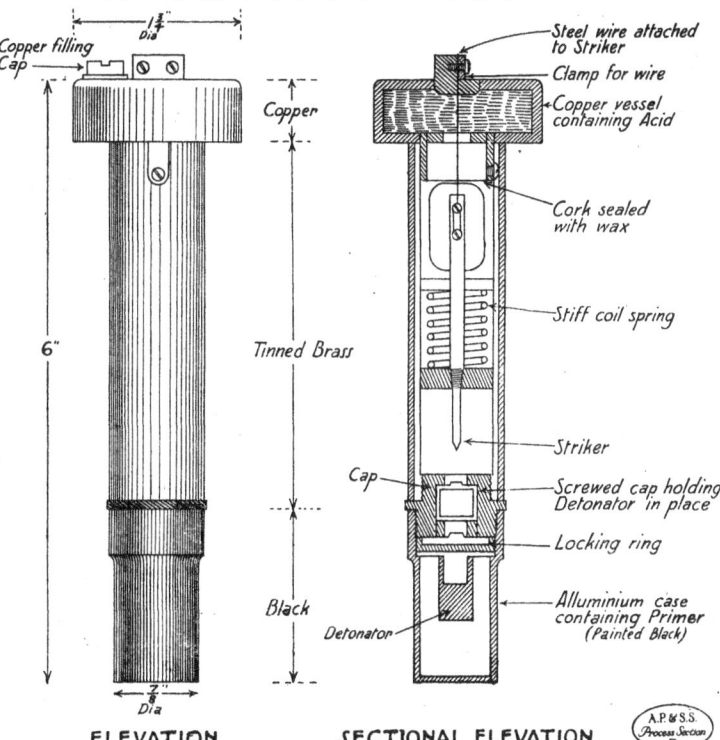

ELEVATION　　**SECTIONAL ELEVATION**

The device shown above *is exceedingly dangerous* and explodes automatically after a certain length of time, owing to the chemical action of the acid severing the piece of fine wire, and thus releasing the striker.

The device should be handled as little as possible after it has been removed from a charge, as it is *liable to explode at any moment.* It should be carried horizontally at arm's length, holding it by the copper head, with the other end away from the body, and buried at least one foot deep or thrown into a well.

SKETCH 1.

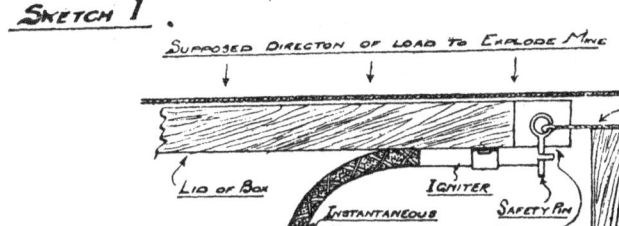

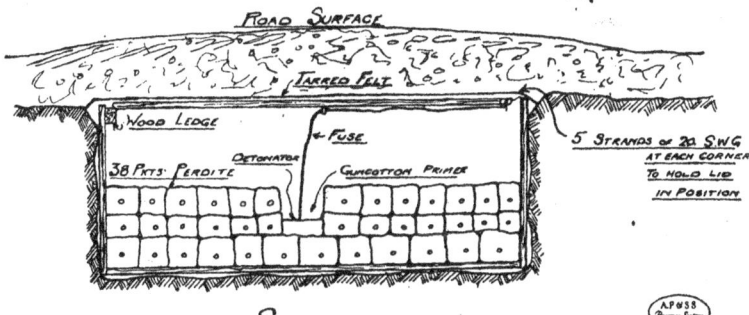

— SECTION —

SKETCH 2.

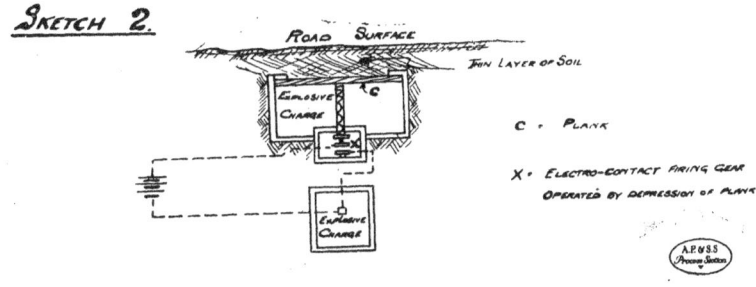

C = PLANK

X = ELECTRO-CONTACT FIRING GEAR
OPERATED BY DEPRESSION OF PLANK

Sketch 3.

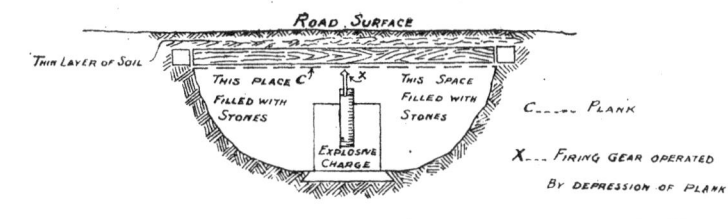

C ------ Plank

X --- Firing gear operated by depression of plank

Sketch 4

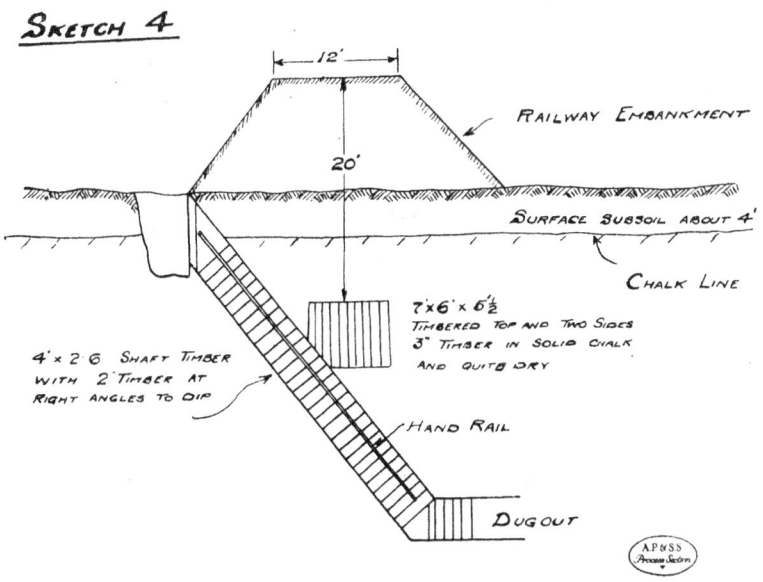

Sketch 5

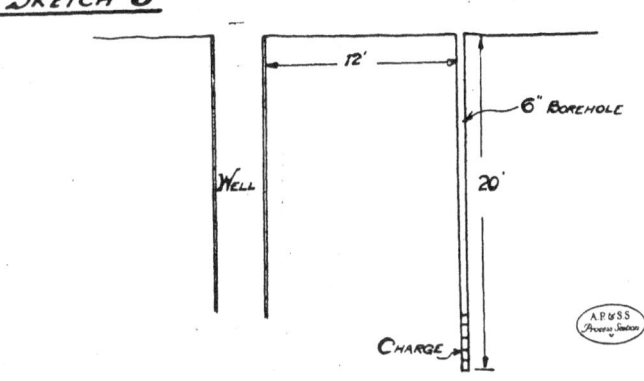

Sketch 6

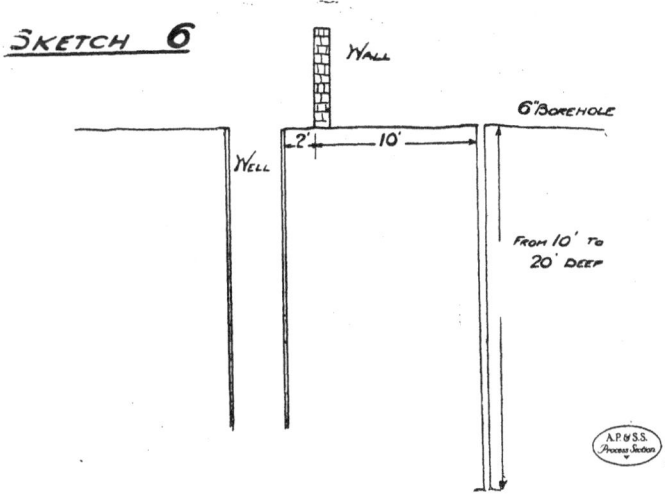

40/W.O./6903.

AMENDMENTS TO S.S. 163.

"HINTS ON RECONNAISSANCE FOR MINES AND LAND MINES IN THE AREA EVACUATED BY THE GERMANS."

(ISSUED BY THE GENERAL STAFF.)

September, 1918.

On page 2.—DUG-OUTS.

After para. 5, *insert* "Gas shells have latterly been used to a considerable extent. They are usually placed at the foot of the incline and fired by time fuze when the dug-out is evacuated. Their employment with trip wire arrangements and delay action fuzes may be anticipated."

On page 5.—RAILWAYS.

Line 6—*Delete* "Trap charges" and *substitute* "Delay action mines."

On page 6.—GENERAL.

In sub-para. 1, *after* "stick-bombs" *add* "or trench mortar bombs."

In sub-para. 5, *after* "charges" *add* "The body of a British soldier was found to have a grenade attached to the wrist in such a manner that it would explode if the body was incautiously moved."

Appendix.

THE 1917 GERMAN LONG DELAY ACTION FUZE FOR DEMOLITION PURPOSES.

(From German documents.)

This fuze screwed into an ordinary H.E. shell, is intended for destroying guns, battery positions, dug-outs, etc., or to blow up ammunition dumps abandoned to the enemy.

Description.—The mechanism is the same as that of the "German Automatic Detonating Device," described in Ia/31532 of the 5th April, 1917, that is, it depends on the action of a corrosive liquid eating through a wire. The fuze contains a detonator and an exploder charge.

It differs in appearance from the "Device," which was about 6 inches long and 1 inch in diameter, with a head $1\frac{3}{4}$ inches in diameter, and was painted black, in being exactly similar outside to the Gr. Z. 04 gun fuze ($8\frac{1}{4}$ inches long and over $1\frac{1}{2}$ inches in diameter, usually used with the 10-cm. gun, 15-cm. and 21-cm. howitzers, and not used in any field artillery shell); with the difference that the gaine of the demolition fuze is painted red and that of the gun fuze lacquered blue.

Means of recognition.—The fuze when in a shell, being marked like the gun fuze Gr. Z. 04, cannot, as the gaine is concealed, be distinguished from an ordinary fuze.

All German shell, however, are issued fuzed. To prevent the fuze working loose, its lower edge is stabbed with a centre punch (usually in four or six places) into corresponding nicks in the top of the shell; or else, as in gas shell, the fuze is set in cement. The absence, therefore, of punch marks on the edge of the fuze, and of cement, may give an indication of the presence of a demolition fuze.

[P.T.O.

Removal.—The fuze can be screwed in by hand, and, therefore, may be removable in the same manner, but can be removed by an adjustable spanner, as there are two flats on it.

If the fuze has been stabbed, a small hand drill will also be required.

Delay.—The fuze, by using different tubes of corrosive liquid, may be set for 1, 2, 24 or 72 hours' delay.

Packing.—The fuzes are packed two together in a special box. On the label in red is:—

Lgz. Z. 17. Nicht verfeuern, nur fur besondere Zwecke.

(1917 pattern long delay action fuzes. Not to be fired; only to be used for special purposes.)

The liquids employed to ensure these "delays" are pale blue in colour, and are issued in glass tubes about ⅜ inch in diameter and 6 inches long. The tube is drawn out and sealed at each end, and is marked either 1, 2, 24 or 72.

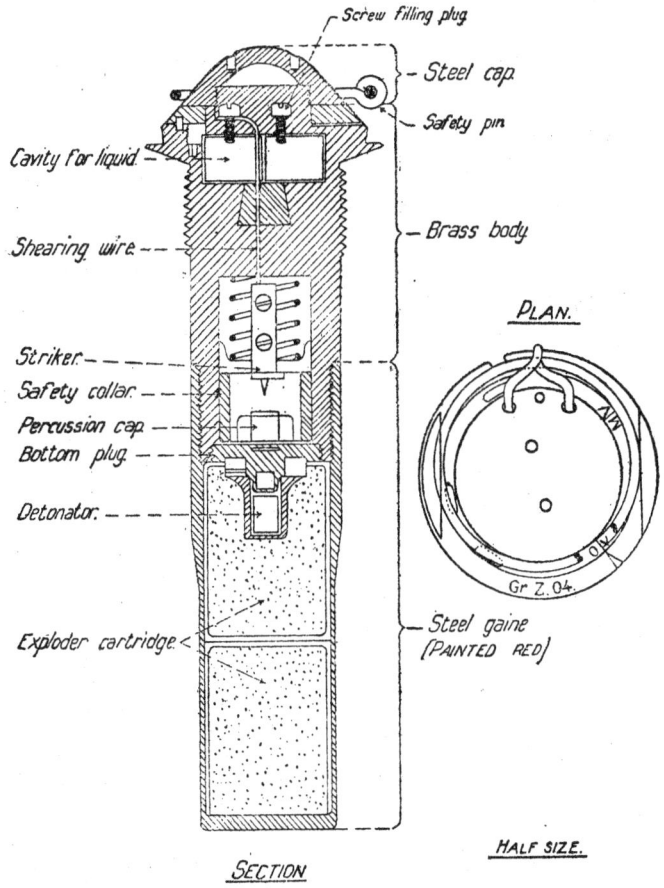

SECTION   HALF SIZE.